PROBABILITY DISTRIBUTIONS

V. Rothschild
N. Logothetis

JOHN WILEY & SONS

New York • Chichester • Brisbane • Toronto • Singapore

Library of Congress Cataloging-in-Publication Data

Rothschild, V.
   Probability distributions.

   Includes index.
   1. Distribution (Probability theory)
I. Logothetis, N.   II. Title.
QA273.6.R67   1986   519.2   85-20308
ISBN 0-471-83814-4

Printed in the United States of America

10 9 8 7 6 5 4 3 2 1

to Pat, Harry and Bridget

PROBABILITY DISTRIBUTIONS

CONTENTS

INTRODUCTION

"Probability distributions" (or just "distributions") is an omnibus phrase covering frequency functions, probability functions, probability mass functions, probability densities, density functions and densities. It is not synonymous with "distribution functions" nor "cumulative distribution functions", $F(x)$.

Students often ask their instructors—or themselves—the question: What does $f(x;n,p) = \binom{n}{x} p^x q^{n-x}$ look like? How will it change if x, n, or p are varied? What about its properties and applications? Although graphical representations of all the distributions in this booklet exist somewhere, most of them, for example in *Distributions in Statistics* by N.L. Johnson & S. Kotz (Houghton Mifflin Co & John Wiley & Sons, Inc.)—expensive, lengthy and out of print—there is no inexpensive work in print in which the common probability distributions are graphically depicted; and it is believed that a useful purpose is served by making available such a compilation, including brief notes on the properties and applications of the distributions and the relationships, if any, between them.

The word "common" needs some comment. The thirty or so distributions included in this booklet, though varying somewhat in sophistication, are believed to be the most common ones. It would have been possible to add other distributions such as the Circular Normal distribution, the Inverse Gaussian distribution, the truncated Poisson distributions or mixtures. Instead, some blank pages are included at the end for such additions, should their need arise.

This booklet contains many symbols such as $E(X)$, $V(X)$, μ, μ_2, μ_3 and μ_4 which are not defined in the text. The Appendix on p. 64, however, provides relevant information. Nevertheless, it should be read in conjuction with a standard statistics texbook such as Hoel, P.G., 1984 *Introduction to Mathematical Statistics* 5th Ed., John Wiley & Sons, New York, or Johnson, R., & Bhattacharyya, G., 1985, *Statistics: Principles and Methods,* John Wiley & Sons, New York.

The distributions described in this booklet are mathematical models. In practice, however, one is almost invariably concerned with samples whose size may be quite small, with concomitant sampling errors as well as those which automatically occur when measurements are made. A small table of such measurements is given in Appendix 2, together with the associated histogram and the theoretical distribution (Normal, p. 26) appropriate to the histogram. What has been said above applies to many, but not all, of the theoretical distributions described in this booklet.

No knowledge of calculus is required to read this booklet because the small number of places where it is needed can, if necessary, be ignored.

For those mainly interested in "what does the distribution look like and how will it change if x, n or p are varied?", all odd numbered pages can be ignored.

A number of the graphs in the following pages were adapted from Johnson & Kotz (already mentioned); Mood, A.M. *et al.,* 1974, *Introduction to the Theory of Statistics,* McGraw-Hill International Book Co., and Mosteller, F. *et al.,* 1970, *Probability with statistical applications,* Addison-Wesley Publishing Co. Inc.

The Bivariate Normal distributions on p. 30 and that on the cover were kindly supplied by the Marconi Research Centre, Chelmsford, England.

V.R.
N.L. 1985

DISCRETE UNIFORM DISTRIBUTION

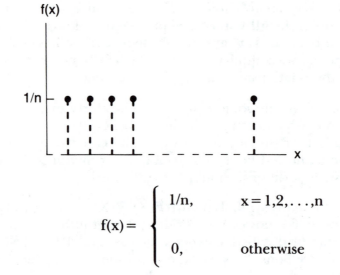

$$f(x) = \begin{cases} 1/n, & x = 1, 2, \ldots, n \\ \\ 0, & \text{otherwise} \end{cases}$$

$$E(X) = \frac{n+1}{2}; \; V(X) = \frac{n^2 - 1}{12}$$

$$\mu_3 = 0$$

$$\mu_4 = (n^2 - 1)(3n^2 - 7)/240$$

$$\psi(t)* = \sum_{i=1}^{n} \frac{1}{n} e^{it}$$

---

*See Appendix 1 for definitions of $\psi(t)$ (moment generating function), $E(X)$ (Expectation), $V(X)$ (Variance), and the moments μ_3 and μ_4.

DISCRETE UNIFORM DISTRIBUTION

PROPERTIES

The distribution of n events, each with probability $P(X=x)=\frac{1}{n}$, when $x=1,2,\ldots,n$, is the Discrete Uniform distribution.

APPLICATION

When a six-sided unbiased die is rolled, the probability of 1 or 2 or . . . or 6 being face uppermost is 1/6 so that $\sum\limits_{i=1}^{6}{}^{*}\frac{1}{n_i}=1$ as required.

      The Discrete Uniform distribution (of any type) is often used, as is the Continuous Uniform distribution (p. 22), for generating random numbers appropriate to any discrete or continuous distribution. The method is always based on obtaining one or more real numbers uniformly distributed between 0 and 1 and then applying a suitable transformation.

RELATIONSHIPS WITH OTHER DISTRIBUTIONS

(a), The discrete analogue of the Continuous Uniform or Rectangular distribution; (b), if, in the Hypergeometric distribution (p. 18), $a=b=-1$ and n can take on any real value, positive or negative, the Discrete Uniform distribution $f(x)=(n+1)^{-1}$, $x=0,2,\ldots,n$ is obtained.

---

*See Appendix 1 for definition of Σ.

BERNOULLI DISTRIBUTION

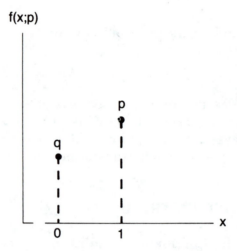

$$f(x;p) = \begin{cases} p^x q^{1-x}, & x = 0,1; \; q = 1-p \\ \\ 0, & \text{otherwise} \end{cases}$$

$$\mathbf{E}(X) = p; \; \mathbf{V}(X) = pq$$

$$\mu_r' = p, \; r \geq 1$$

$$\psi(t) = q + pe^t$$

BERNOULLI DISTRIBUTION

PROPERTIES

This distribution is appropriate to a *single* trial, the outcome of which is Success or Failure—Yes or No—Heads or Tails—Cured or Died—or, symbolically, 1 or 0. In the Bernoulli distribution opposite, the probability, p, of Success is about .61 and, therefore of Failure, q, about .39.

APPLICATIONS

This distribution applies to any experiment consisting of trials of the types mentioned above. Another example is the probability of a new-born baby being a boy or a girl (hermaphrodites excluded). As most people know, the probability of a boy is approximately but not exactly 1/2 (1.05 boys to 1.00 girls *born*).

RELATIONSHIPS WITH OTHER DISTRIBUTIONS

The Bernoulli distribution is a special case of the Binomial distribution (p. 8) when n = 1 and of the Multinomial distribution (p. 10) when k = 2 and n = 1.

BINOMIAL DISTRIBUTIONS

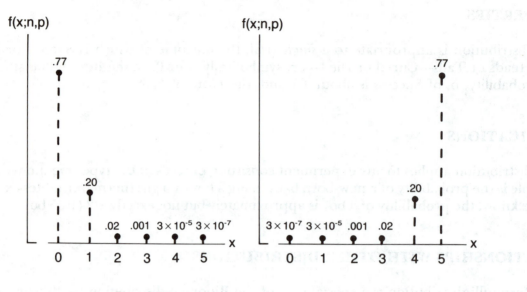

n=5,p=0.05

n=5,p=0.95

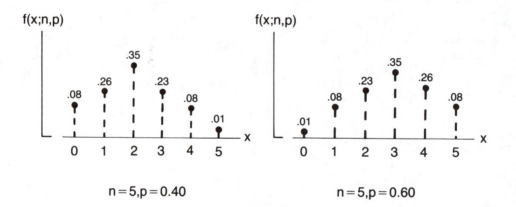

n=5,p=0.40

n=5,p=0.60

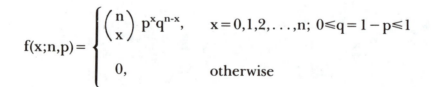

$$f(x;n,p) = \begin{cases} \dbinom{n}{x} p^x q^{n-x}, & x=0,1,2,\ldots,n;\ 0 \leq q = 1-p \leq 1 \\\\ 0, & \text{otherwise} \end{cases}$$

$$E(X) = np;\ V(X) = npq$$

$$\mu_3 = npq(q-p)$$

$$\mu_4 = 3n^2p^2q^2 + npq(1-6pq)$$

$$\psi(t) = (q + pe^t)^n$$

BINOMIAL DISTRIBUTION

PROPERTIES

If the probability of observing an event E is p for each of n independent Bernoulli trials (p. 6), the probability that E will be observed in exactly x of these trials is given by the Binomial distribution. This distribution derives its name from the Binomial Theorem, $(a+b)^n = a^n + \binom{n}{1} a^{n-1}b + \binom{n}{2} a^{n-2}b^2 + \ldots + \binom{n}{n-1} ab^{n-1} + b^n$, in which, if q and p are substituted for a and b, each term corresponds with the Binomial distribution for a particular value of x, e.g. when x = 1, f(1;n,p) or $P(X=1) = \binom{n}{1} pq^{n-1}$.

APPLICATIONS

A classical example of the event E, referred to above, is the outcome of tossing a coin, which may or may not be biased ($p \neq 1/2$ or $p = 1/2$) n times and determining the probability of, say, exactly x = 4 heads in n = 6 tosses. If p = 1/2, $f(4;6,1/2) = \binom{6}{4} (1/2)^4 (1/2)^2 = 0.\underset{2344}{\cancel{4688}}$. Alternatively, a doctor may be interested in the probability that 7 out of 10 patients recovered by chance (i.e. p = 1/2) after being treated with a particular drug, that is 7 successes in 10 trials. The answer is $0.\underset{\text{0.1172}}{\cancel{7031}}$. The Binomial distribution can be used in tests of the goodness of fit type, helping to determine whether it is reasonable to believe that the proportions (or frequencies) observed in a sample could have been derived from a population having a particular value of p.

RELATIONSHIPS WITH OTHER DISTRIBUTIONS

(a), When $n \geq 50$ and $np < 5$, the Binomial distribution can be approximated by a Poisson distribution (p. 16) with mean $\lambda = np$; (b), similarly, the Normal distribution (p. 26) with $\mu = np$ and $\sigma = \sqrt{npq}$ is a good approximation to the Binomial distribution provided np and nq are greater than 5; (c), finally, the Hypergeometric distribution (p. 18) reduces to the Binomial distribution as N, the number of elements in the Hypergeometric population, tends to infinity or is large compared with n, the sample size. In these circumstances, replacement and non-replacement tend to become indistinguishable.

MULTINOMIAL DISTRIBUTION

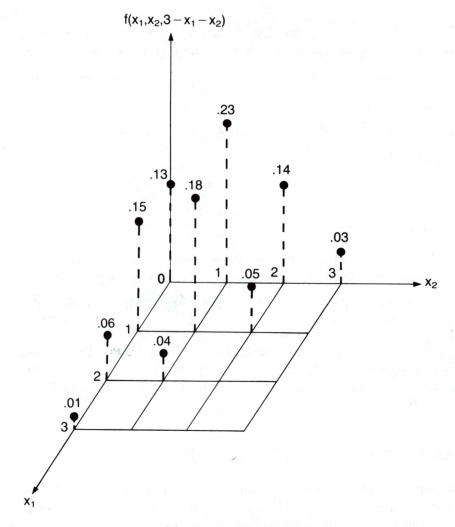

$$n = 3, k = 3, p_1 = 0.2, p_2 = 0.3$$

$$f(x_1, \ldots, x_k; n, p_i) = \frac{n! \prod_{i=1}^{k} p_i^{x_i}}{\prod_{i=1}^{k} x_i!}, \qquad x_i = 0, \ldots, n; \ \sum_{i=1}^{k} x_i = n; \ \sum_{i=1}^{k} p_i = 1; \ p_i = 1 - q_i; \ i = 1, 2, \ldots, k$$

$$\mathbf{E}(X_i) = np_i; \ \mathbf{V}(X_i) = np_i q_i$$

$$\mu'_{r_1 \cdots r_k} = n(n-1) \ldots \left(n - \sum_{i=1}^{k} r_i + 1 \right) \prod_{i=1}^{k} p_i^{r_i}$$

$$\psi(t_1, \ldots, t_k) = \left(\sum_{i=1}^{k} p_i e^{t_i} \right)^n$$

MULTINOMIAL DISTRIBUTION

PROPERTIES

This distribution is an extension of the Binomial distribution (p. 8), in which successive independent trials have as their outcomes Success or Failure, in that an experiment consisting of n trials has k discrete and mutually exclusive possible outcomes $E_1, E_2, \ldots, E_k$ with probabilities $p_1, p_2, \ldots, p_k$. The probability that E_1 will occur x_1 times, E_2 x_2 times and so on where $\sum_{i=1}^{k} x_i = n$, can be calculated, by analogy with a derivation of the Binomial distribution, as follows: the probability of obtaining the sequence $\overbrace{E_1, \ldots, E_1}^{x_1}$; $\overbrace{E_2, \ldots, E_2}^{x_2}$; $\ldots$; $\overbrace{E_k, \ldots, E_k}^{x_k}$ is $\prod_{i=1}^{k}{}^{*} p_i^{x_i}$ (since the trials are independent). Every arrangement of the above set of Es has this same probability of occurrence. The number of arrangements is the number of permutations of n objects of which x_1 are alike, x_2 are alike etc., which is $n! / \prod_{i=1}^{k} x_i!$. As all the arrangements are the mutually exclusive ways in which the required event can occur and as each has the probability given above, multiplication of the latter by the number of permutations produces the multinomial distribution which incidentally is the general term in the expansion of $\left(\sum_{i=1}^{k} p_i \right)^n$.

APPLICATIONS

A classical example concerns an urn containing, say, 5 green (g) balls, 4 black (b) balls and 3 yellow (y) balls. A ball is chosen at random, its colour is noted and it is returned to the urn. What is the probability that if 6 balls are chosen as above, 3 will be green, 2 will be black and 1 will be yellow? The answer is: $\mathbf{P}(3g, 2b, 1y) = \frac{6!}{3!2!1!} \left(\frac{5}{12} \right)^3 \left(\frac{4}{12} \right)^2 \left(\frac{3}{12} \right)^1 = 625/5184$. This urn model has applications to natural phenomena such as the kinetic theory of gases in classical physics.

RELATIONSHIPS WITH OTHER DISTRIBUTIONS

(a), When k = 2 it reduces to the Binomial distribution; and (b), when k = 2 and n = 1, to the Bernoulli distribution (p. 6).

---

*See Appendix 1 for definition of Π.

GEOMETRIC DISTRIBUTIONS

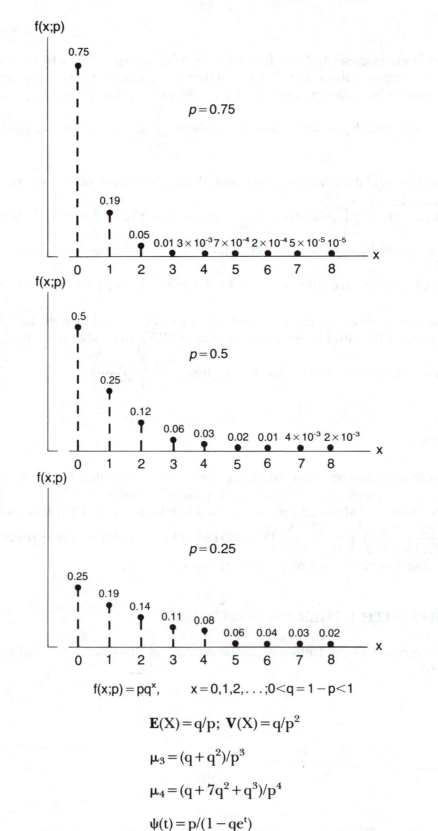

$$f(x;p) = pq^x, \qquad x = 0,1,2,\ldots; 0 < q = 1 - p < 1$$

$$\mathbf{E}(X) = q/p; \quad \mathbf{V}(X) = q/p^2$$

$$\mu_3 = (q + q^2)/p^3$$

$$\mu_4 = (q + 7q^2 + q^3)/p^4$$

$$\psi(t) = p/(1 - qe^t)$$

GEOMETRIC DISTRIBUTION

PROPERTIES

In a sequence of independent Bernoulli trials (p. 6), the number of failures before the first success occurs has a Geometric distribution with parameter p.

Suppose that there is a long sequence of failures in a series of independent Bernoulli trials with p = 0.5. Contrary to the intuitive feelings of some gamblers this has no effect on future trial outcomes, a phenomenon called *the memoryless* or *Markov property* of the distribution. Mathematically, for any non-negative integers s and t, $P(X = s + t | X \geqslant s) = P(t)$.

The frequencies of this distribution always fall off in geometric progression as the value of X increases: hence the name.

APPLICATIONS

The probability that 5 tosses of an unbiased coin are needed to get one Head (i.e. 4 consecutive Tails) is $f(4; 0.5) = 0.5(0.5)^4 = 0.0313$. Instead of being formulated as $f(x; p) = pq^x$, $x = 0, 1, 2, \ldots$, where X is the number of failures before a success, the Geometric distribution may be expressed as $f(x; p) = pq^{x-1}$, $x = 1, 2, \ldots$, in which case X is the number of tosses up to and including the first success. The Geometric distribution is quite often used in meteorological models of weather cycles.

RELATIONSHIPS WITH OTHER DISTRIBUTIONS

The discrete analogue of the Exponential distribution (p. 36), and a special case of the Negative Binomial distribution (p. 14) because, if $X_1, X_2, \ldots, X_r$ are independent and identically distributed random variables, each having a Geometric distribution with parameter p, $\sum_{i=1}^{r} X_i$ will have a Negative Binomial distribution with parameters p and r.

NEGATIVE BINOMIAL DISTRIBUTIONS

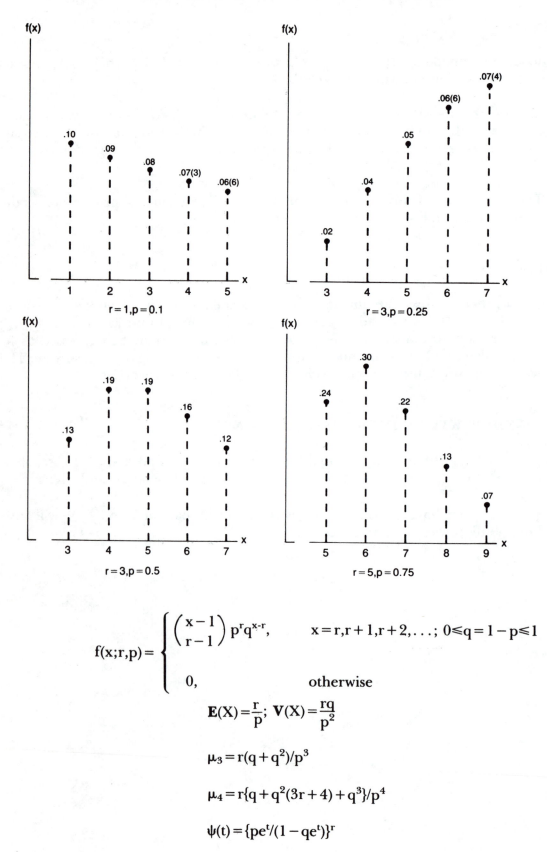

$$f(x;r,p) = \begin{cases} \begin{pmatrix} x-1 \\ r-1 \end{pmatrix} p^r q^{x-r}, & x=r,r+1,r+2,\ldots; \ 0 \leq q = 1-p \leq 1 \\ \\ 0, & \text{otherwise} \end{cases}$$

$$\mathbf{E}(X) = \frac{r}{p}; \ \mathbf{V}(X) = \frac{rq}{p^2}$$

$$\mu_3 = r(q+q^2)/p^3$$

$$\mu_4 = r\{q+q^2(3r+4)+q^3\}/p^4$$

$$\psi(t) = \{pe^t/(1-qe^t)\}^r$$

NEGATIVE BINOMIAL DISTRIBUTION

PROPERTIES

If independent Bernoulli trials (p. 6) with constant p are repeated X times until a fixed number r of successes has occurred, X has a Negative Binomial distribution. This distribution can be considered as the sum of a series of random variables $X_1 + X_2 + \ldots + X_r$ each with the same Geometric distribution (p. 12), i.e.

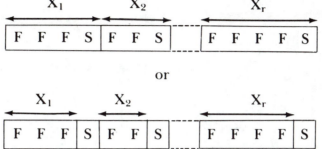

The second diagram refers to the alternative form of the distribution: $f(x;r,p,) = \binom{x+r-1}{x} p^r q^x$, $x = 0,1,2,\ldots$, in which X is the number of trials beyond r, i.e. the number of failures needed for r successes referred to as S($=$ Success), above.

APPLICATIONS

When an unbiased coin is tossed, the probability of a second head on the fourth toss is $f(4;2,0.5) = P(X=4) = \binom{3}{1} (0.5)^2 (0.5)^2 = 0.1875$. The distribution finds uses in accident statistics and as a substitute for a Poisson distribution (p. 16) when strict independence requirements are not satisfied. It can describe consumer expenditure and demand for often bought products.

Suppose that the number of individuals in a single group is a random variable with Logarithmic distribution $f(x;p) = -q^x/x\log p$ (p. 20). Further, that the number of such groups is also a random variable whose distribution is $f(y) = e^{-\lambda}\lambda^y/y!$ (Poisson). The product of these two random variables defines another random variable representing the total number of individuals which has a Negative Binomial distribution $f(z) = \binom{z-1}{r-1} p_N^r q_N^{z-r}$ where $r = -\lambda/\log p$, $p_N = q/p$ and $q_N = 1 - p_N$.

RELATIONSHIPS WITH OTHER DISTRIBUTIONS

(a), When $r = 1$ this distribution becomes the Geometric distribution; (b), it is related to the Binomial distribution (p. 8) by $f(x;r,p) = f(r-1;x-1,p) \cdot p$ where the right-hand function f is the binomial probability of $r-1$ successes in $x-1$ trials; (c), there is a general relationship between the Negative Binomial, the Poisson and the Binomial distributions in that all three are generated from the expansion of $\{(1+\omega)-\omega\}^{-m}$. For the Negative Binomial case $\omega > 0$ and $m > 0$; for the Binomial case $-1 < \omega < 0$ and $m < 0$; and for the Poisson case $\omega \to 0$ and $m \to 0$ while $\omega m = \lambda$; (d), the Logarithmic distribution (p. 20) is a limiting form of the Negative Binomial distribution when the latter is truncated by exclusion of the first term corresponding to $x = r$.

POISSON DISTRIBUTIONS

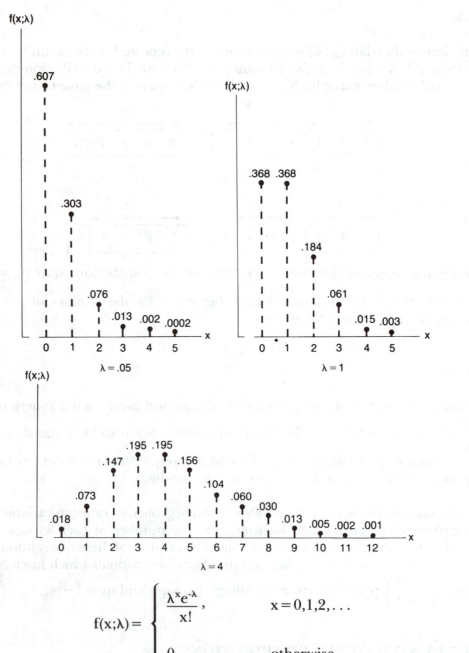

$$f(x;\lambda) = \begin{cases} \dfrac{\lambda^x e^{-\lambda}}{x!}, & x = 0,1,2,\ldots \\[3mm] 0, & \text{otherwise} \end{cases}$$

$$E(X) = \lambda; \ V(X) = \lambda$$

$$\mu_3 = \lambda$$

$$\mu_4 = \lambda + 3\lambda^2$$

$$\psi(t) = e^{\lambda(e^t - 1)}$$

POISSON DISTRIBUTION

PROPERTIES

A random variable X has a Poisson distribution if (a), the number of outcomes of a trial occurring in one time interval or specified region is independent of the number occurring in any other non-overlapping time interval or region of space (events occur at random in continuous time or space); (b), the probability of a single outcome during a short time interval or in a small region of space is proportional to the length of the time interval or the size of the region and does not depend on the number of outcomes occurring outside this time interval or region; (c), the probability that more than one outcome will occur in such a short time interval or small region is negligible. Given these conditions, the steps in the mathematical derivation of the Poisson distribution are straightforward and given in several, but not all, statistical textbooks. A further property of the Poisson distribution is that the sum of n independent Poisson variables each with parameter λ_i, $i = 1, 2, \ldots, n$ is also a Poisson variable with parameter $\sum_{i=1}^{n} \lambda_i$.

APPLICATIONS

A typist makes on average 2 mistakes per page. What is the probability that he or she will make 4 mistakes on a page? $f(4; \lambda) = P(X = 4) = 2^4 e^{-2}/4! = 0.0902$. Perhaps the most famous application relates to the number of α particles reaching a given portion of space during time t; but it is also important for blood or sperm counts in both of which, after suitable dilution of the original material, the number of cells in a known volume of diluent is counted, after time has been given for their sedimentation.

RELATIONSHIPS WITH OTHER DISTRIBUTIONS

(a), If, in a Binomial distribution (p. 8), $n \geqslant 50$ and p is close to 0 (i.e. $np < 5$), the distribution is closely approximated by a Poisson distribution with $\lambda = np$. More formally, the conditions are $n \to \infty$, $p \to 0$, but $np = \lambda$ remains constant; (b), on the other hand, the limiting distribution of the standardized Poisson variable $(X - \lambda)/\lambda^{1/2}$, where X has a Poisson distribution and when $\lambda \to \infty$, the standard Normal distribution (p. 26) is obtained; (c), if X has a Poisson distribution, $P(x \leqslant X) = P(Y > 2\lambda)$ where Y has a $\chi^2_{(2x+2)}$ distribution (p. 40); (d), finally, in a Poisson process the intervals between successive events have independent identical Exponential distributions (p. 36). Consequently, the number of events in a specified time interval has a Poisson distribution.

HYPERGEOMETRIC DISTRIBUTIONS

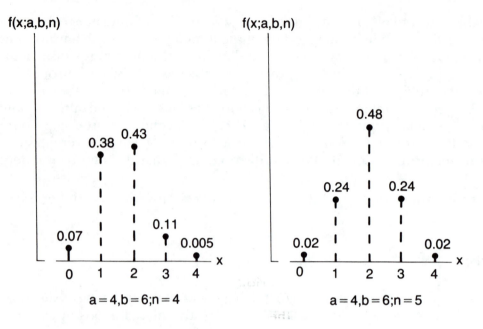

$$f(x;a,b,n) = \begin{cases} \dfrac{\dbinom{a}{x}\dbinom{b}{n-x}}{\dbinom{a+b}{n}}, & x=0,1,2,\ldots,n;\ 0\le x\le(a+b),0<n\le(a+b) \\[2em] 0, & \text{otherwise} \end{cases}$$

$$\mathbf{E}(X) = \frac{na}{a+b};\ \mathbf{V}(X) = \frac{nab(a+b-n)}{(a+b)^2(a+b-1)}$$

$$\mu_3' = \frac{na}{a+b}\left\{ \frac{(n-1)(a-1)}{(a+b-1)}\left[\frac{(n-2)(a-2)}{(a+b-2)} + 3 \right] + 1 \right\}$$

$$\mu_4' = \frac{na}{a+b}\left\{ \frac{(n-1)(a-1)}{(a+b-1)}\left[\frac{(n-2)(a-2)}{(a+b-2)}\left(\frac{(n-3)(a-3)}{(a+b-3)} + 6 \right) + 7 \right] + 1 \right\}$$

$\psi(t)$, of no use

HYPERGEOMETRIC DISTRIBUTION

PROPERTIES

The important feature of this distribution is that it refers to sampling from a population *without* replacement of the sample. Suppose that an urn contains a red balls and b black balls and that an experiment consisting of n trials is performed, a ball being chosen from the urn at random, its colour noted and the ball then being returned to the urn. If X is the random variable number of red balls chosen in n trials, the Binomial distribution (p. 8) requires that the probability of exactly x successes is

$$f(x;n,p)= \binom{n}{x} \frac{a^{x}b^{n-x}}{(a+b)^{n}}$$ because $p=a/(a+b)$ and $q=1-p=b/(a+b)$. If, however, sampling is without

replacement, this Binomial distribution becomes the Hypergeometric distribution. When the population being sampled is infinite, the two distributions are identical.

APPLICATIONS

In industrial quality control, runs of size N are subjected to sampling inspection. Defective items in the run, the number of which N_1 is unknown, are "red balls". A sample of size n is taken and the number a of defective items is noted. The Hypergeometric distribution enables inferences to be made from this information about the probable magnitude of N_1. The precise method of estimation is beyond the scope of this booklet. An interesting application of the Hypergeometric distribution is in the estimation of the size of animal populations from "capture-recapture" data. After catching a sample of particular animals, tagging them and letting them return to the population from which they were sampled, another sample is caught and, according to the number of tagged animals in this second sample, the size of the whole population can be estimated.

RELATIONSHIPS WITH OTHER DISTRIBUTIONS

(a), This distribution can be approximated by a Binomial distribution with parameters n and $p=a/(a+b)$; (b), since the latter distribution can itself be approximated by a Poisson (p. 16) or Normal (p. 26) distribution, there is a corresponding Poisson or Normal approximation to the Hypergeometric distribution, when n is large, $\geqslant 50$, and $a/(a+b)$ is small, $<.1$, in the Poisson case, and when n is large and $a/(a+b)$ is not small, ≈ 0.5, in the Normal case; (c), a Generalized Hypergeometric distribution is obtained when a, b and n are allowed to have any real, positive or negative values. In the special case when $a=b=-1$, the Discrete Uniform distribution (p. 4) is obtained.

LOGARITHMIC DISTRIBUTIONS

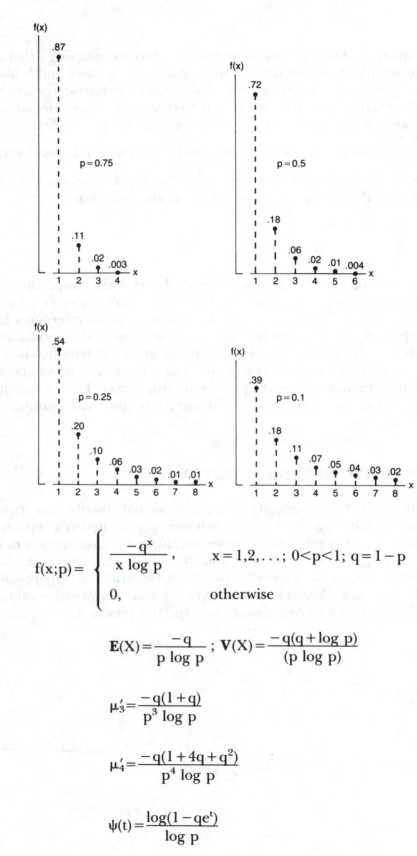

$$f(x;p) = \begin{cases} \dfrac{-q^x}{x \log p}, & x=1,2,\ldots;\ 0<p<1;\ q=1-p \\[2ex] 0, & \text{otherwise} \end{cases}$$

$$E(X) = \frac{-q}{p \log p}\ ;\ V(X) = \frac{-q(q+\log p)}{(p \log p)}$$

$$\mu_3' = \frac{-q(1+q)}{p^3 \log p}$$

$$\mu_4' = \frac{-q(1+4q+q^2)}{p^4 \log p}$$

$$\psi(t) = \frac{\log(1-qe^t)}{\log p}$$

LOGARITHMIC DISTRIBUTION

PROPERTIES

If the first value of x in the Negative Binomial distribution (p. 14) is missing so that $f(x;r,p) = p^r rq, p^r r(r+1)q^2/2!, \ldots, x = r+1, r+2, \ldots$, and as $\Sigma f(x;r,p) = 1 - p^r$ in these circumstances, this distribution tends to the following form of the Logarithmic distribution as r tends to zero:

$$\frac{-1}{\log p}\left(q, \frac{q^2}{2}, \frac{q^3}{3}, \ldots\right), \qquad x = 1, 2, \ldots$$

In other words the frequency at integral values $k \geq 1$ is the coefficient of t^k in the expansion of $\{\log(1-qt)\}/\{\log p\}$. These distributions have long positive tails so that for large values of X, the shape of the tail is similar to that of a Geometric distribution (p. 12).

APPLICATIONS

This distribution has applications in connection with the spatial frequency distributions of plants and animals and with the growth of populations. It is also used to represent the distribution of products bought in a specified period of time.

RELATIONSHIPS WITH OTHER DISTRIBUTIONS

(a), This distribution can be derived from the Negative Binomial distribution (p. 14) when r in $f(x;r,p)$ tends to low (<0.1) non-integral values; (b), there is, also, the following relationship between the Logarithmic, Poisson (p. 16) and Negative Binomial distributions: if the number of "groups" of individuals has a Poisson distribution with $E(X) = \lambda$ and the number of individuals per "group" has a Logarithmic distribution, the total number of individuals has an approximately Negative Binomial distribution with parameters $r = \lambda/(-\log p)$ and $p_{N.B.} = q/p$; (c), a discrete Lognormal distribution (p. 32) can be obtained by assigning a Logarithmic distribution to the parameters of a Poisson distribution.

CONTINUOUS UNIFORM OR RECTANGULAR DISTRIBUTION

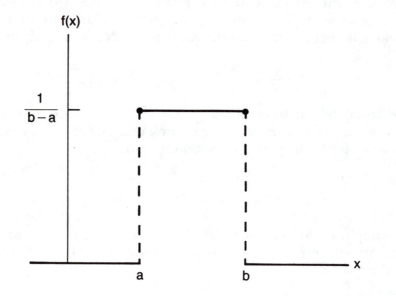

$$f(x) = \begin{cases} \dfrac{1}{b-a}, & a < x < b \\[2ex] 0, & \text{otherwise} \end{cases}$$

$$E(X) = \frac{a+b}{2} \; ; \; V(X) = \frac{(b-a)^2}{12}$$

$$\mu_3 = 0$$

$$\mu_4 = \frac{(b-a)^4}{80}$$

$$\psi(t) = \frac{e^{tb} - e^{ta}}{t(b-a)}$$

CONTINUOUS UNIFORM OR RECTANGULAR DISTRIBUTION

PROPERTIES

This distribution is often used as a model for the errors due to rounding off, up or down, when recording measurements. If the lengths of some objects are measured to the nearest millimeter, one of the objects whose length is recorded as 63mm has in reality a length which is equally likely to have any value between 62.5 and 63.5mm. A reasonable rounding off model in this case, therefore, requires that the errors are uniformly distributed on the open interval $(-0.5, 0.5)$.

APPLICATIONS

The Continuous Uniform distribution has many uses in making mathematical models of physical, biological and social phenomena. A mathematical application is given above. A further mathematical application is the use of this distribution in the derivation of Sheppard's correction which adjusts the values of sample moments for the grouping of data.

RELATIONSHIPS WITH OTHER DISTRIBUTIONS

(a), This distribution, the continuous analogue of the Discrete Uniform distribution (p. 4), is a special case of the Beta distribution (p. 50) when $\alpha = \beta = 1$; (b), if X has a Continuous Uniform distribution written in the equivalent form $f(x) = (2h)^{-1}$, $m - h \leqslant x \leqslant m + h$; $h > 0$, and if $m = h = \frac{1}{2}$, $Z = -2\log X$ has an Exponential distribution (p. 36) $f(z) = \frac{1}{2} e^{-z/2}$, $z > 0$, i.e. Z is distributed as χ^2_2 (p. 40); and (c), if S_2 denotes the sum of two independent random variables each with the above distribution with $m = h = \frac{1}{2}$, the distribution of $\frac{1}{2} S_2$ is symmetrical and Triangular (p. 24).

TRIANGULAR DISTRIBUTION

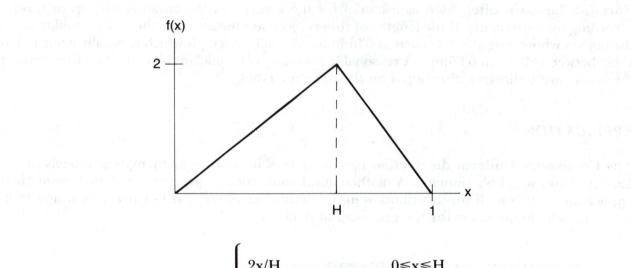

$$f(x) = \begin{cases} 2x/H, & 0 \leq x \leq H \\ 2(1-x)/(1-H), & H \leq x \leq 1 \end{cases}$$

$$\mathbf{E}(X) = (1/3)(1+H); \quad \mathbf{V}(X) = (1/18)(1-H+H^2)$$

$$\mu_r' = \frac{2(H^r + H^{r-1} + \ldots + H + 1)}{(r+1)(r+2)}, \quad r \geq 1$$

$\psi(t)$, of no use

TRIANGULAR DISTRIBUTION

PROPERTIES

This distribution's name derives from the fact that when plotted, its shape is triangular. When symmetrical, i.e. when $H = 1/2$, it is sometimes called a Tine distribution.

APPLICATIONS

It can be used as a crude approximation to the Normal distributions (p. 26). In fact, in its symmetrical form it can equally well be an approximation to any other symmetrical continuous distribution.

RELATIONSHIPS WITH OTHER DISTRIBUTIONS

The arithmetic mean of two independent random variables each with the same Continuous Uniform distribution (p. 22) has a Triangular distribution.

NORMAL (OR GAUSSIAN) DISTRIBUTIONS

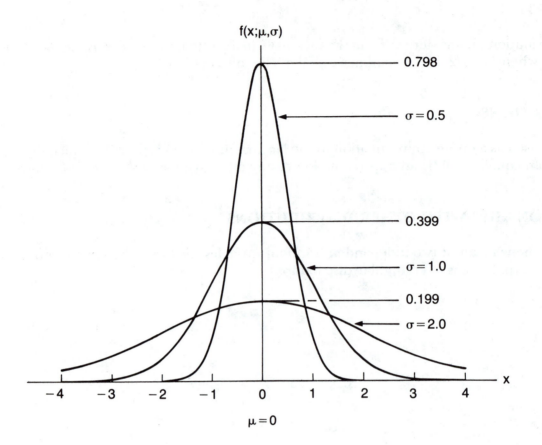

$$f(x;\mu,\sigma) = \frac{1}{\sqrt{2\pi}\,\sigma}\, e^{\frac{-(x-\mu)^2}{2\sigma^2}}, \qquad -\infty < \mu < \infty; \sigma > 0$$

$$\mathbf{E}(X) = \mu; \quad \mathbf{V}(X) = \sigma^2$$

$$\mu_r = \begin{cases} 0, & r \text{ odd} \\[2ex] \dfrac{r!\sigma^r}{(r/2)!2^{r/2}}, & r \text{ even} \end{cases}$$

$$\psi(t) = e^{\mu t + (1/2)\sigma^2 t^2}$$

Note If $\mu = 0$ and $\sigma = 1$, the Standard Normal distribution $\mathbf{N}(0,1) = \dfrac{1}{\sqrt{2\pi}}\, e^{-(1/2)x^2}$, is obtained.

NORMAL DISTRIBUTION

DERIVATION

Consider a sequence of independent but not necessarily identically distributed random variables $X_1, X_2, \ldots, X_n$ and assume that $E(X_i) = \mu_i$ and $V(X_i) = \sigma_i^2$, $i = 1, 2, \ldots, n$. Let

$$Y_n = \left(\sum_{i=1}^{n} X_i - \sum_{i=1}^{n} \mu_i \right) \Bigg/ \sqrt{\sum_{i=1}^{n} \sigma_i^2}$$

so that $E(Y_n) = 0$ and $V(Y_n) = 1$. If $E(|X_i - \mu_i|^3) < \infty$, $i = 1, 2, \ldots, n$ and

$$\lim_{n \to \infty} \left[\sum_{i=1}^{n} E(|X_i - \mu_i|^3) \right] \Bigg/ \left(\sum_{i=1}^{n} \sigma_i^2 \right)^{3/2} = 0,$$

for any fixed number x $\lim_{n \to \infty} P(Y_n \leq x) = \Phi(x) = \dfrac{1}{\sqrt{2\pi}} \int_{-\infty}^{x} e^{-1/2 u^2} \, du$, $-\infty < x < \infty$. This Standard Normal distribution, with $\mu = 0$ and $\sigma^2 = 1$, is often written $N(0,1)$.

This is an exposition of the *Central Limit Theorem*. In words, the standardised sum of n independent random variables is asymptotically normal.

APPLICATIONS

Many measurements such as those of the heights and weights of human beings, of the diameters of machined parts, of the lengths of tobacco leaves, of measurement errors and of IQ scores are *approximately* normally distributed. One cause of "approximately" is that the tails of the observed distributions contain more probability, i.e. are fatter, than those of the Normal distribution.

RELATIONSHIPS WITH OTHER DISTRIBUTIONS.

(a), There are Normal approximations to the Binomial (p. 8), Poisson (p. 16), χ^2 (p. 40) and the standard Student's t (p. 54) distributions; (b), on the other hand, a Normal distribution can be approximated by a Lognormal distribution (p. 32) if the coefficient of variation, $\sqrt{V(X)}/\mu$, in the Normal has a small absolute value, <0.25; (c), a particular form of the Logistic distribution (p. 58), that is when $\alpha = 0$ and $\beta = \sqrt{3}/\pi$, is very close to the Normal distribution; (d), the Weibull distribution (p. 46) with $b \approx 3.25$ is almost identical with the Normal distribution. If Z_i, $i = 1, 2, \ldots, n$ have a standard Normal distribution and are independent: (e), $\sum_{i=1}^{n} Z_i^2$ has a χ_n^2 distribution; (f), Z_1/Z_2 has Student's t distribution with $\nu = 1$ and

therefore the Cauchy distribution (p. 52); (g), $\left(\sum_{i=1}^{m} Z_i^2 / m \right) \Bigg/ \left(\sum_{i=1}^{n} Z_i^2 / n \right)$ has an F distribution (p. 56) with m and n degrees of freedom.

STANDARD NORMAL DISTRIBUTIONS
TRUNCATED AT a = 1, −1; 1.5, −1.5; 2, −2; and ∞, −∞

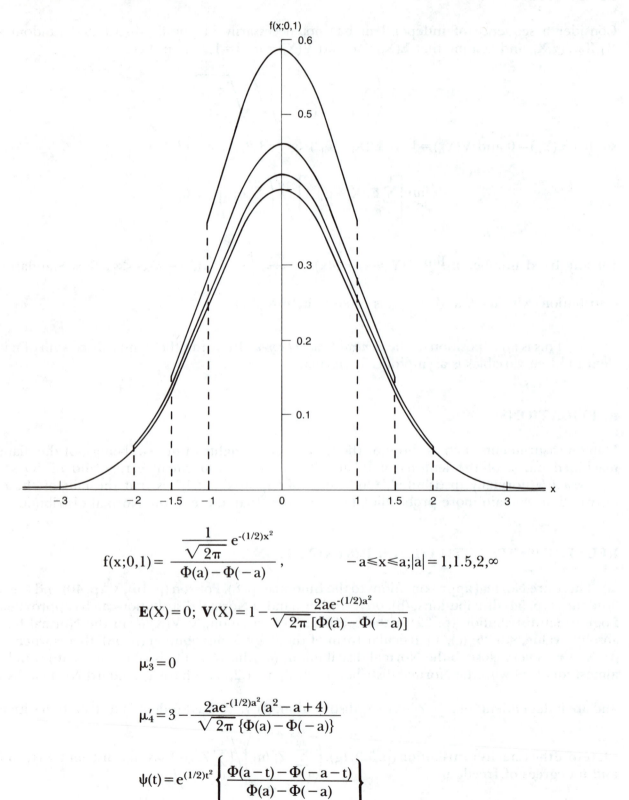

$$f(x;0,1) = \frac{\dfrac{1}{\sqrt{2\pi}}\, e^{-(1/2)x^2}}{\Phi(a) - \Phi(-a)}, \qquad -a \leq x \leq a; |a| = 1, 1.5, 2, \infty$$

$$\mathbf{E}(X) = 0; \quad \mathbf{V}(X) = 1 - \frac{2ae^{-(1/2)a^2}}{\sqrt{2\pi}\,[\Phi(a) - \Phi(-a)]}$$

$$\mu_3' = 0$$

$$\mu_4' = 3 - \frac{2ae^{-(1/2)a^2}(a^2 - a + 4)}{\sqrt{2\pi}\,\{\Phi(a) - \Phi(-a)\}}$$

$$\psi(t) = e^{(1/2)t^2} \left\{ \frac{\Phi(a-t) - \Phi(-a-t)}{\Phi(a) - \Phi(-a)} \right\}$$

TRUNCATED NORMAL DISTRIBUTION

PROPERTIES

The distribution of truncated normal variables (or functions of them) usually cannot be expressed in a mathematically elegant form. When the degree of truncation $\Phi(-a)$ is large or when the degree of truncation is elaborate, for example when it is asymmetric or a number of X intervals are omitted, the distribution bears little resemblance to a Normal distribution (p. 26). Nevertheless it can still approximately fit large samples from experimental data in which measurements are only available for a particular range of the variable.

APPLICATIONS

Selection procedures that sometimes operate with possible truncation effects can be found in life-testing and response-time studies. Such a case could arise in a time-mortality experiment in which the first observation is delayed until a fixed time interval has elapsed so that n_1 sample specimens die before observation begins. The experiment is subsequently terminated at a predetermined time, with n_2 specimens remaining alive. Actual survival times are recorded for the n specimens which die during the whole period of observation. In such an experiment, the logarithm of the survival time can be assumed to be normally distributed, in which case a doubly truncated Normal distribution is obtained.

RELATIONSHIPS WITH OTHER DISTRIBUTIONS

All the relationships with other distributions relevant to the Normal distribution are equally relevant to the Truncated Normal distribution.

BIVARIATE NORMAL DISTRIBUTIONS
WITH VARYING CORRELATION COEFFICIENTS ρ

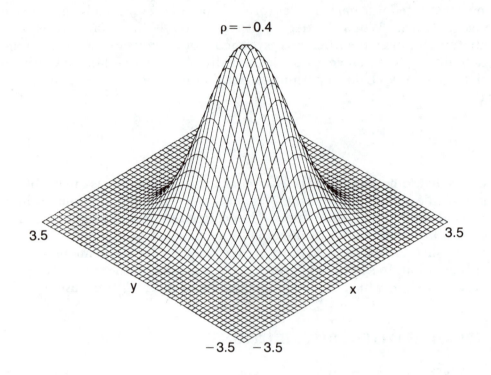

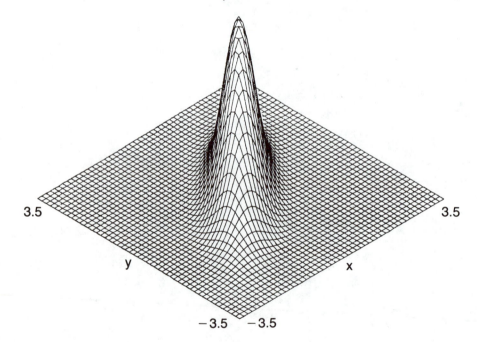

BIVARIATE NORMAL DISTRIBUTIONS

$$f(x,y) = \frac{1}{2\pi(1-\rho^2)^{1/2}\sigma_X\sigma_Y} \exp\left\{-\frac{1}{2(1-\rho^2)}\left[\left(\frac{x-\mu_X}{\sigma_X}\right)^2 - 2\rho\left(\frac{x-\mu_X}{\sigma_X}\right)\left(\frac{y-\mu_Y}{\sigma_Y}\right) + \left(\frac{y-\mu_Y}{\sigma_Y}\right)^2\right]\right\}$$

$$-\infty < x < \infty;\ -\infty < y < \infty; \rho = \frac{Cov(x,y)^*}{\sigma_X\sigma_Y}$$

$$\mathbf{E}(X) = \mu_X;\ \mathbf{V}(X) = \sigma_X^2;\ \mathbf{E}(Y) = \mu_Y;\ \mathbf{V}(Y) = \sigma^2_Y;\ \rho(X,Y) = \rho$$

$$\mu'_{11} = \rho\sigma_1\sigma_2 + \mu_1\mu_2$$
$$\mu'_{12} = 2\rho\sigma_1\sigma_2\mu_2 + \mu_1(\mu_2^2 + \sigma_2^2)$$
$$\mu'_{21} = 2\rho\sigma_1\sigma_2\mu_1 + \mu_2(\mu_1^2 + \sigma_1^2)$$
$$\mu'_{22} = (\mu_1\mu_2 + \sigma_1\sigma_2)^2 + (\sigma_1\mu_2 + \sigma_2\mu_1)^2 + 2\rho^2\sigma_1^2\sigma_2^2$$
$$\psi_{X,Y}(t_1,t_2) = \exp\{t_1\mu_1 + t_2\mu_2 + 1/2(t_1^2\sigma_1^2 + 2\rho t_1t_2\sigma_1\sigma_2 + t_2^2\sigma_2^2)\}$$

PROPERTIES

If X and Y have a Bivariate Normal distribution, their marginal distributions are Normal (p. 26). The regression of either variable on the other is linear, i.e. $Y = aX + b + \epsilon_1$, and $X = cY + d + \epsilon_2$, the random variables ϵ_1 and ϵ_2 having identical distributions. If $\mu_X = \mu_Y = 0$ and $\sigma_X^2 = \sigma_Y^2 = 1$, the distribution of $Y|X = x$ is normal with expected value ρx and standard deviation $\sqrt{1-\rho^2}$. A similar result holds for $X|Y = y$. It is to be noted that two normally distributed variables do not necessarily have a joint Bivariate Normal distribution.

APPLICATIONS

This distribution is often used as an approximation to joint distributions even when the marginal distributions are not exactly Normal. There are applications in biometric data and in artillery fire control in which the distribution is used as an approximation for the deviation from some target on a plane.

RELATIONSHIPS WITH OTHER DISTRIBUTIONS

If the parameters α and β of a Gumbel distribution (p. 60) are estimated by $\bar{\alpha}$ and $\bar{\beta}$, the joint asymptotic standardised distribution of $\bar{\alpha}$ and $\bar{\beta}$ is Bivariate Normal. The marginal distributions of a Bivariate Normal distribution are univariate and Normal. If (X,Y) has a Bivariate Normal distribution with $\sigma_X^2 = \sigma_Y^2 = (2\alpha)^{-1}$ and $\rho = 0$, $R^2 = X^2 + Y^2$, called the Radial error, has a Rayleigh distribution (p. 44) with parameter α. If (X,Y) has a standardised Bivariate Normal distribution, i.e. $\mu_X = \mu_Y = 0$ and $\sigma_X^2 = \sigma_Y^2 = 1$, the ratio X/Y has a Cauchy distribution (p. 52) with density $f(x) = \sqrt{1-\rho^2}/\{\pi(1 - 2\rho x + x^2)\}$.

---

*See Appendix 1 for Definition of Cov(x,y).

LOGNORMAL DISTRIBUTIONS

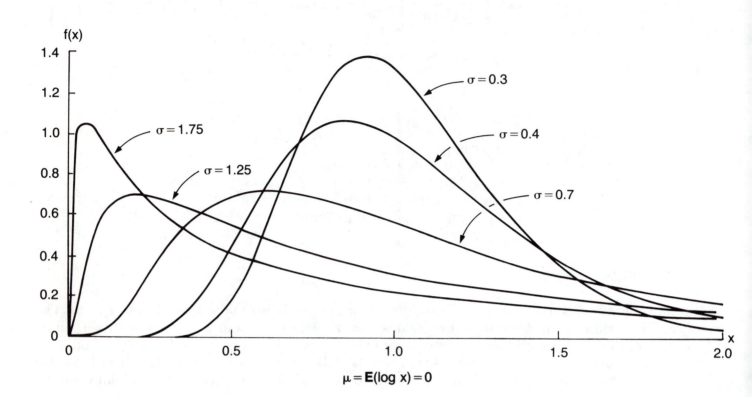

$$\mu = \mathbf{E}(\log x) = 0$$

$$f(x;\mu,\sigma) = \frac{1}{\sqrt{2\pi\sigma^2 x^2}}\, e^{-(\log x-\mu)^2/2\sigma^2}, \qquad 0 < x < \infty;\ -\infty < \mu < \infty;\ \sigma > 0$$

$$\mathbf{E}(X) = e^{\mu + 1/2\sigma^2};\ \mathbf{V}(X) = e^{2\mu}(e^{2\sigma^2} - e^{\sigma^2})$$

$$\mu_r' = e^{1/2 r^2\sigma^2 + r\mu}, r \geq 1$$

$$\psi(t),\ \text{of no use}$$

LOGNORMAL DISTRIBUTION

PROPERTIES

If the random variable X has a Normal distribution (p. 26) and if the values of the random variable Y are related to those of X by the equation $y = e^x$, Y has a Lognormal distribution.

APPLICATION

The Lognormal distribution is widely applied in practical statistical work, e.g. the distribution of particle size in natural aggregates and of economic units, critical dosages in drug applications and the duration of doctors' consultations.

RELATIONSHIPS WITH OTHER DISTRIBUTIONS

(a), If σ (on p. 31) is sufficiently small, <0.2, a Lognormal distribution very similar to the Normal distribution can be constructed; (b), the Lognormal distribution is also a serious competitor of the Weibull distribution (p. 46) for the representation of life-time distributions (manufactured products), and (c), of the Logarithmic distribution (p. 20) when the zero class is omitted and it therefore is in a truncated form.

MAXWELL DISTRIBUTIONS

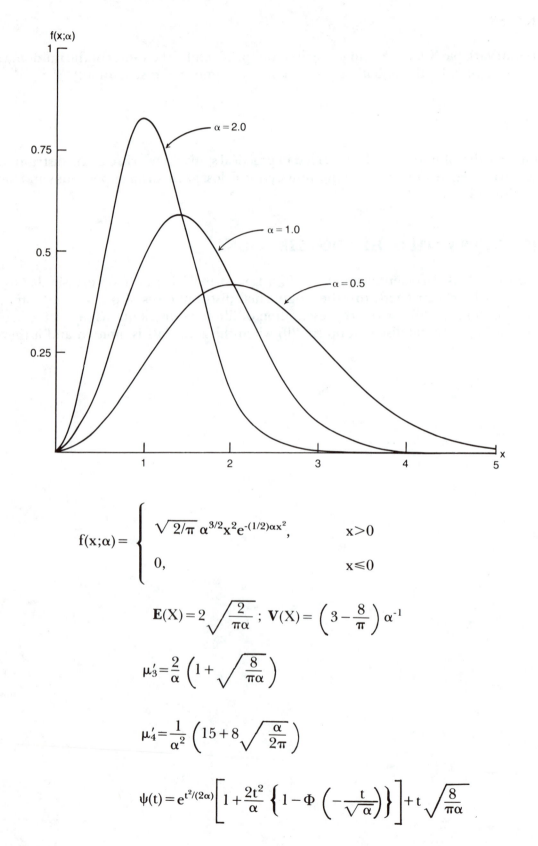

$$f(x;\alpha) = \begin{cases} \sqrt{2/\pi}\ \alpha^{3/2}x^2 e^{-(1/2)\alpha x^2}, & x > 0 \\ 0, & x \leq 0 \end{cases}$$

$$E(X) = 2\sqrt{\frac{2}{\pi\alpha}}\ ; \ V(X) = \left(3 - \frac{8}{\pi}\right)\alpha^{-1}$$

$$\mu_3' = \frac{2}{\alpha}\left(1 + \sqrt{\frac{8}{\pi\alpha}}\right)$$

$$\mu_4' = \frac{1}{\alpha^2}\left(15 + 8\sqrt{\frac{\alpha}{2\pi}}\right)$$

$$\psi(t) = e^{t^2/(2\alpha)}\left[1 + \frac{2t^2}{\alpha}\left\{1 - \Phi\left(-\frac{t}{\sqrt{\alpha}}\right)\right\}\right] + t\sqrt{\frac{8}{\pi\alpha}}$$

MAXWELL DISTRIBUTION

The Maxwell distribution is referred to in some serious statistical textbooks but Kendall & Buckland* say: "A fortunately rare expression for a chi-squared distribution [see p. 40 in this booklet] with three degrees of freedom, based presumably on Clark Maxwell's discussion of the energy of particles moving at random in three dimensions." In fact, the Maxwell distribution becomes $a\sqrt{\chi_3^2}$ distribution when α, see opposite, equals 1.

According to Maxwell, the speed of a molecule of an ideal gas can be considered as the outcome of a stochastic experiment having as its sample space the set of non-negative real numbers $[0,\infty)$, and having a probability distribution defined by $f(v;h,m)=kv^2\exp(-hmv^2), v>0$, where k, h and m are constants (see opposite). k is determined so that $\int_\Omega f(v)dv = 1$. The probability that the speed of a molecule falls in a set E of non-negative and real numbers is $P(E)=\int_E f(v)dv$.

APPLICATIONS

The Maxwell distributions may be used to describe the speed of a particle moving in 3-dimensional space such that its movements along the three coordinate axes are Normally distributed independent random variables (p. 26) with mean zero and variance α^{-1}. If the particle moves in 2-dimensional space its speed is better described by the Rayleigh distribution (p. 44). Knowing the distribution of the speed v, the Maxwell distribution can be used to find the distribution of the particle's kinetic energy which is related to v by the formula $E=1/2mv^2$, where m is the mass of the particle. In such a case, the parameter $\alpha=m/kT$ where T is the absolute temperature and k is Boltzmann's constant (1.38062×10^{-23} joule/degree K).

The Maxwell distribution can also be used to describe the distribution of $\sqrt{e_1^2+e_2^2+e_3^2}$ where e_1, e_2, and e_3 are the measurement errors in the position coordinates of a particle in 3-dimensional space.

*Kendall, Sir M.G. & Buckland, W.R., 1982, *A dictionary of statistical terms*, p. 120, Longman Group Ltd., London and New York.

EXPONENTIAL DISTRIBUTION

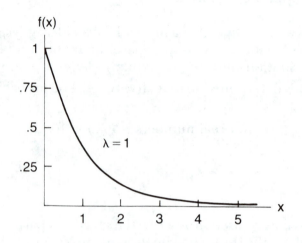

$$f(x;\lambda) = \begin{cases} \lambda e^{-\lambda x}, & x \geq 0; \ \lambda > 0 \\ \\ 0, & x < 0 \end{cases}$$

$$\mathbf{E}(X) = 1/\lambda; \ \mathbf{V}(X) = 1/\lambda^2$$

$$\mu_r' = \frac{r!}{\lambda^r}, \qquad r \geq 1$$

$$\psi(t) = \frac{\lambda}{\lambda - t}, \qquad t < \lambda$$

EXPONENTIAL DISTRIBUTION

PROPERTIES

If Y is a random variable with some arbitrary continuous distribution function $F(y), x = -\log\{1 - F(y)\}$ has a standard Exponential distribution, that is with $\lambda = 1$. This property is of special importance in the theory of Order Statistics since it permits study of the Order Statistics of samples from a population with any continuous distribution through the Order Statistics of exponential variables.

Another important characteristic is the *memoryless property* (see p. 13) which means that the future lifetime of some object has the same distribution, irrespective of the time for which it has existed. Formally, if X is a continuous positive random variable representing lifetime, $P(X \leq x_0 + x | X > x_0) = P(X \leq x)$, $x_0 > 0$ and $x > 0$. A further, analogous characteristic is the "constant expected residual life" property:

$$E(X - k | X \geq k) = E(X), \ k > 0$$

APPLICATIONS

This distribution is widely used to describe events recurring at random in time. The lifetime or life characteristic can often be usefully represented by an exponential random variable, with a relatively simple associated theory. Among the most frequent applications are those in the field of life testing. It is also used to get approximate solutions to difficult distribution problems.

RELATIONSHIPS WITH OTHER DISTRIBUTIONS

(a), The continuous analogue of the Geometric distribution (p. 12); (b), it is also a special case of the Gamma distribution (p. 42) when α in the latter equals 1; (c), the sum of n exponential variables each with the same parameter λ is a Gamma variable with parameters $\alpha = n$ and $\beta = \lambda$; (d), when $\lambda = 1/2$ in the Exponential distribution it becomes a χ_2^2 distribution (p. 40); (e), but if a random variable X is such that $Y = (X - \theta)^b$ has an Exponential distribution, X becomes a Weibull distribution with shape parameter b (p. 46); (f), if N is a random variable defined by the inequalities $\sum_{i=1}^{N} T_i \leq \tau < \sum_{i=1}^{N+1} T_i$ where $T_i, i = 1, 2, \ldots$ are independent exponentially distributed random variables, N has a Poisson distribution (p. 16) with parameter $\lambda\tau$; (g), the difference between two exponentially distributed random variables gives rise to the Double Exponential distribution (p. 38).

DOUBLE EXPONENTIAL DISTRIBUTION

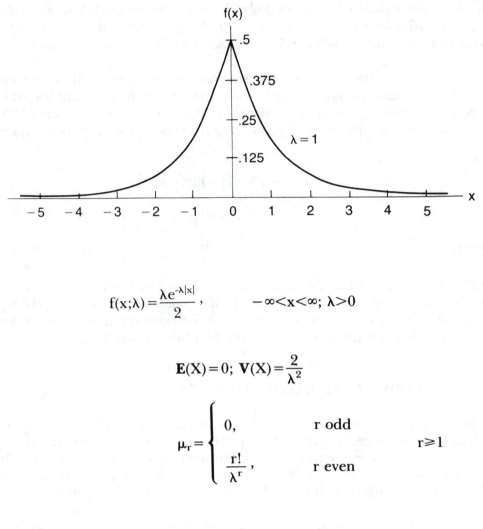

$$f(x;\lambda) = \frac{\lambda e^{-\lambda|x|}}{2}, \qquad -\infty < x < \infty; \; \lambda > 0$$

$$\mathbf{E}(X) = 0; \; \mathbf{V}(X) = \frac{2}{\lambda^2}$$

$$\mu_r = \begin{cases} 0, & r \text{ odd} \\[2mm] \dfrac{r!}{\lambda^r}, & r \text{ even} \end{cases} \qquad r \geq 1$$

$$\psi(t) = \frac{\lambda^2}{\lambda^2 - t^2}, \qquad t < \lambda$$

DOUBLE EXPONENTIAL

PROPERTIES

This is the type of distribution whose likelihood function is maximised by setting the location parameter equal to the median of the observed values of an odd number of independent identically distributed random variables. It is also known as the "first law of Laplace". When $\lambda = 1$ the form sometimes called "Poisson's first law of error" is obtained. The Exponential distribution (p. 36) can be thought of as a Double Exponential "folded" about $x = 0$.

APPLICATIONS

This distribution sometimes provides a convenient alternative to the Normal distribution (p. 26), if symmetry can be assumed.

RELATIONSHIPS WITH OTHER DISTRIBUTIONS

If X has a Double Exponential distribution, $|X|$ is distributed exponentially, or as $(2\lambda)^{-1}\chi_2^2$ (p. 40). If Z_1, Z_2, Z_3, and Z_4 are independent standard Normal variables $Z_1Z_4 - Z_2Z_3$ has a Double Exponential distribution with $\lambda = 1/2$. The Cauchy distribution with $\alpha = 0$ and $\lambda = 1$ (see p. 52) and hence Student's t distribution (p. 54) with one degree of freedom are connected with the Double Exponential distribution $f(x;\lambda)$ by the following relationship:

$$f(x;1) = 1/2 \int_{-\infty}^{\infty} f(v;0,1)e^{ivx}dv$$

where $i = \sqrt{-1}$.

χ² DISTRIBUTIONS
FOR VARIOUS DEGREES OF FREEDOM ν

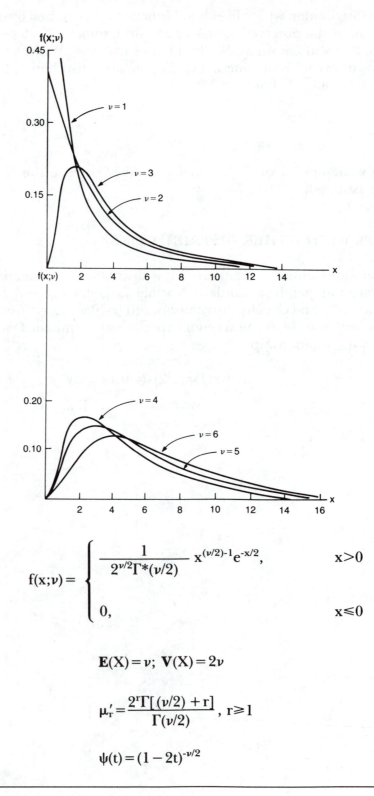

$$f(x;\nu) = \begin{cases} \dfrac{1}{2^{\nu/2}\Gamma^*(\nu/2)}\, x^{(\nu/2)-1}e^{-x/2}, & x > 0 \\[3em] 0, & x \leq 0 \end{cases}$$

$$E(X) = \nu; \quad V(X) = 2\nu$$

$$\mu_r' = \frac{2^r \Gamma[(\nu/2)+r]}{\Gamma(\nu/2)}, \ r \geq 1$$

$$\psi(t) = (1-2t)^{-\nu/2}$$

---

*See Appendix 1 for properties of Γ[ ].

χ^2 DISTRIBUTION

PROPERTIES

The χ^2 distribution appears naturally in the theory associated with normally distributed random variables as the distribution of the sum of squares of independent standard Normal variables (p. 26). In fact, as ν tends to infinity, the χ_ν^2 distribution tends to normality (although rather slowly), with all the associated Normal properties. For $\nu>2$, the distribution has a mode at $\nu-2$. For $\nu=2$, the distribution is J-shaped with the maximum ordinate at zero. For $0<\nu<2$ the distribution is again J-shaped but with an infinite ordinate at the origin.

APPLICATIONS

The distribution is primarily used as an approximation for the χ^2 statistic which is valuable for various tests of significance, the most prominent being that of the goodness of fit comparison tests between observed and hypothetical frequencies falling into specified classes. Alternatively, it is used for comparisons between observed and hypothetical variances in Normal samples, and to test the independence of two variables.

RELATIONSHIPS WITH OTHER DISTRIBUTIONS

(a), This distribution is a special case of the Gamma distribution (p. 42); (b), it is also the limiting distribution, as n→∞, of the discrete random variable $\sum_{i=1}^{k} (N_i - np_i)^2 (np_i)^{-1}$ when $N_1, N_2, \ldots, N_k$ have a joint Multinomial distribution (p. 10) with parameters $n, p_1, p_2, \ldots, p_k$; (c), if Z_i, $i=1,2,\ldots,\nu$ are standard Normal random variables, $\sum_{i=1}^{\nu} Z_i^2$ has a χ_ν^2 distribution; (d), the ratio $(\chi_{\nu_1}^2/\nu_1)/(\chi_{\nu_2}^2/\nu_2)$ has an F_{ν_1,ν_2} distribution (p. 56); (e), if Y has a standard Continuous Uniform distribution, $f(y;0,1)=1, 0<y<1$ (p. 22), $-2\log Y$ is distributed as χ_2^2.

GAMMA DISTRIBUTIONS

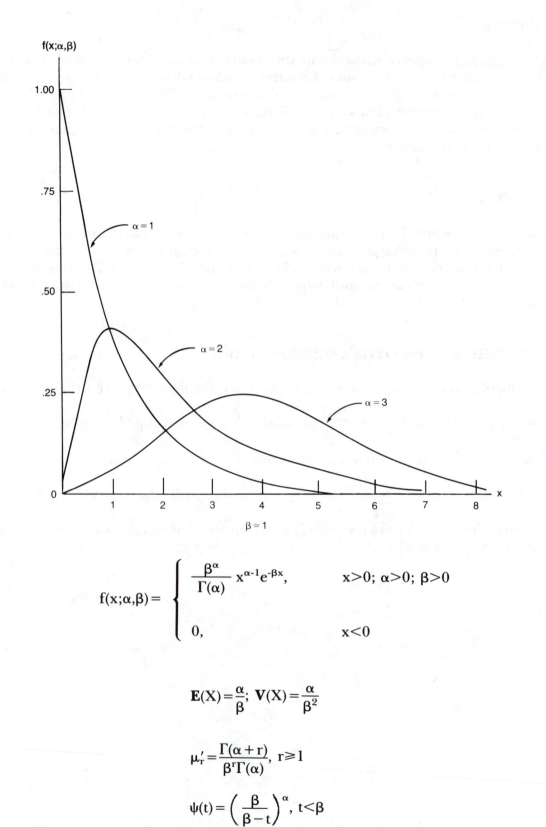

$$f(x;\alpha,\beta) = \begin{cases} \dfrac{\beta^{\alpha}}{\Gamma(\alpha)}\, x^{\alpha-1} e^{-\beta x}, & x>0;\ \alpha>0;\ \beta>0 \\[2em] 0, & x<0 \end{cases}$$

$$\mathbf{E}(X) = \frac{\alpha}{\beta};\ \ \mathbf{V}(X) = \frac{\alpha}{\beta^{2}}$$

$$\mu'_{r} = \frac{\Gamma(\alpha+r)}{\beta^{r}\Gamma(\alpha)},\ r\geqslant 1$$

$$\psi(t) = \left(\frac{\beta}{\beta-t}\right)^{\alpha},\ t<\beta$$

GAMMA DISTRIBUTION

PROPERTIES

One of the most important characteristics of this distribution is its "reproductive property": If X_1 and X_2 are independent random variables with Gamma distributions $f(X_1;\alpha_1,\beta)$ and $f(X_2;\alpha_2,\beta)$, $(X_1 + X_2)$ also has a Gamma distribution with parameters $(\alpha_1 + \alpha_2)$ and β. As the parameter $(\alpha_1 + \alpha_2)$ increases, the shape of the distribution becomes similar to that of the Normal distribution (p. 26).

APPLICATIONS

The Gamma distribution provides useful representations of many physical phenomena. It has been used to make realistic adjustments to Exponential distributions (p. 36) in representing lifetimes in "life-testing", providing a rather flexible skewed density over the positive range. It would, for example, be a useful failure-time model for a system under continuous maintenance if it experiences some wear or degradation initially, but reaches a stable state of repair as time goes on. There are many other applications, for example in meteorological processes, inventory theory, insurance risk theory, economics and queuing theory.

RELATIONSHIPS WITH OTHER DISTRIBUTIONS

(a), Gamma distributions have as special cases the χ^2 distribution (p. 40) when $\alpha = \nu/2$ and $\beta = 1/2$, and the Exponential distribution when $\alpha = 1$ and $\beta = \lambda$; (b), they share with Lognormal distributions (p. 32) the property of closely mimicking a Normal distribution when α is sufficiently large, >15; (c), if X_1 and X_2 are independent standard Gamma random variables, the random variable $X_1/(X_1 + X_2)$ has a standard Beta distribution (p. 50).

RAYLEIGH DISTRIBUTIONS

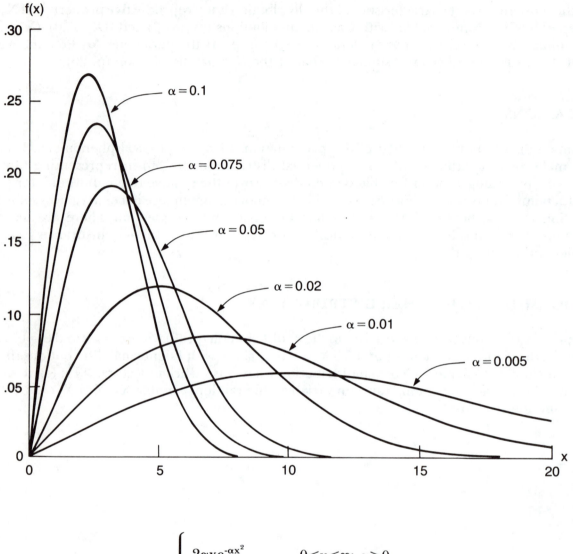

$$f(x;\alpha) = \begin{cases} 2\alpha x e^{-\alpha x^2}, & 0 < x < \infty; \ \alpha > 0 \\ \\ 0, & x \leq 0 \end{cases}$$

$$E(X) = \frac{1}{2}\sqrt{\frac{\pi}{\alpha}}; \ V(X) = \frac{1}{\alpha}\left(1 - \frac{\pi}{4}\right)$$

$$\mu_r' = \alpha^{-r/2}\Gamma\left(1 + \frac{r}{2}\right), \ r \geq 1$$

$\psi(t)$, of no use

RAYLEIGH DISTRIBUTION

PROPERTIES

This distribution may be derived from the distribution of the Radial error R where $R^2 = X^2 + Y^2$ and (X,Y) has a Bivariate Normal distribution with $\sigma_X^2 = \sigma_Y^2 = (2\alpha)^{-1}$ and $\rho = 0$ (p. 31).

APPLICATIONS

Mixtures of Rayleigh distributions are used to describe the hourly median power and instantaneous power of received radio signals. In a generalised form this distribution is used in mathematical physics, particularly in communication theory. There is also a use in generating random numbers.

RELATIONSHIPS WITH OTHER DISTRIBUTIONS

χ_2 (p. 40) (i.e. $\sqrt{\chi_{\nu=2}^2}$) has a Rayleigh distribution with $\alpha = 1/2$. A general form of this distribution is obtained from $\sum_{i=1}^{\nu} (z_i + c_i)^2$ when z_i, $i = 1,2,\ldots,\nu$ are independent standard Normal variables (p. 26) and

c_i, $i = 1,2,\ldots$, are constants. A special case is the χ_ν^2 distribution when $\sum_{i=1}^{\nu} c_i^2 = 0$. The generalised Rayleigh distribution is also called the *non-central* χ_ν^2 distribution with non-centrality parameter $\lambda = \sum_{i=1}^{\nu} c_i^2$.

WEIBULL DISTRIBUTIONS

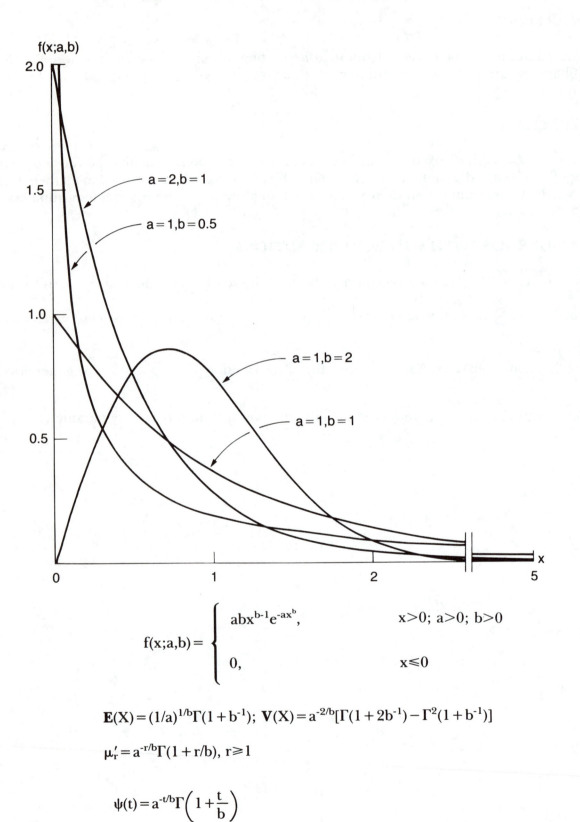

$$f(x;a,b) = \begin{cases} abx^{b-1}e^{-ax^{b}}, & x>0;\ a>0;\ b>0 \\ \\ 0, & x\leq 0 \end{cases}$$

$$\mathbf{E}(X) = (1/a)^{1/b}\Gamma(1+b^{-1});\ \mathbf{V}(X) = a^{-2/b}[\Gamma(1+2b^{-1}) - \Gamma^{2}(1+b^{-1})]$$

$$\mu_r' = a^{-r/b}\Gamma(1+r/b),\ r\geq 1$$

$$\psi(t) = a^{-t/b}\Gamma\left(1+\frac{t}{b}\right)$$

WEIBULL DISTRIBUTION

PROPERTIES

The distribution provides extra flexibility when an Exponential distribution (p. 36) might be adequate, especially when conditions of strict randomness are not satisfied. It describes data arising from life and fatigue tests.

APPLICATIONS

This distribution is often used to represent the distribution of breaking strengths of materials in reliability and quality control tests.

RELATIONSHIPS WITH OTHER DISTRIBUTIONS

(a), If X has this distribution, X^b generates a random variable with an Exponential distribution e^{-x^b}; (b), if X has a Weibull distribution with a variable scale parameter a, which is assumed to have a Gamma distribution (p. 42), X^b will have a Pareto distribution (p. 62); (c), for values of b of about 3.6, the Weibull distribution resembles a Normal distribution (p. 26); (d), when X has a Weibull distribution, the probability distribution of $y = \log(X^{-b}/\alpha)$ is $f(y) = e^{-y}e^{-e^{-y}}$ which is a special case of the Gumbel distribution (p. 60); (e), finally when b is large, >5, the Weibull distribution can be approximated by a Gumbel distribution.

ARC-SINE DISTRIBUTION

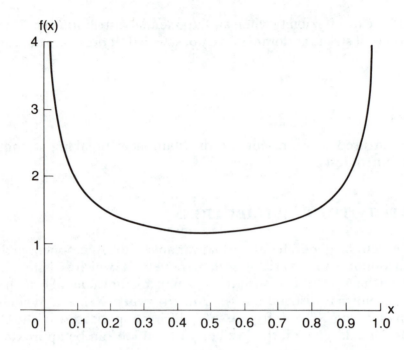

$$f(x) = \frac{1}{\pi\sqrt{x(1-x)}}, \qquad 0<x<1$$

$$\mathbf{E}(X) = 0.5; \ \mathbf{V}(X) = 0.125$$

$$\mu'_r = \frac{\Gamma(r+1/2)}{r!\sqrt{\pi}}, \qquad r \geqslant 1$$

$\psi(t)$, of no use

Notes (i), $\mathbf{V}(X) = \dfrac{\alpha\beta}{(\alpha+\beta)^2(\alpha+\beta+1)}$ when $\alpha = \beta = 0.5$

(ii), $F(x) = \dfrac{2}{\pi}\sin^{-1}\sqrt{x}$

ARC-SINE DISTRIBUTION

PROPERTIES

If a particle moves in a straight line along the x-axis by steps of unit length, starting from zero, and if it is equally likely that a step will be to the left or right, the fraction of time spent on the positive part of the x-axis has an Arc-sine distribution as the number of steps n→∞.

APPLICATIONS

The Arc-sine distribution is connected with many paradoxes. One example is that in a long coin tossing experiment, contrary to intuition, one of the players is quite likely to remain almost the whole time on the winning, e.g. Heads, side. In 20 coin tosses, for example, the probability that the lead never changes from one player, say Heads, to the other is as high as about 0.35.

RELATIONSHIP WITH OTHER DISTRIBUTIONS

This distribution is a special case of the Beta distribution (p. 50) when $\alpha = \beta = 1/2$.

BETA DISTRIBUTIONS

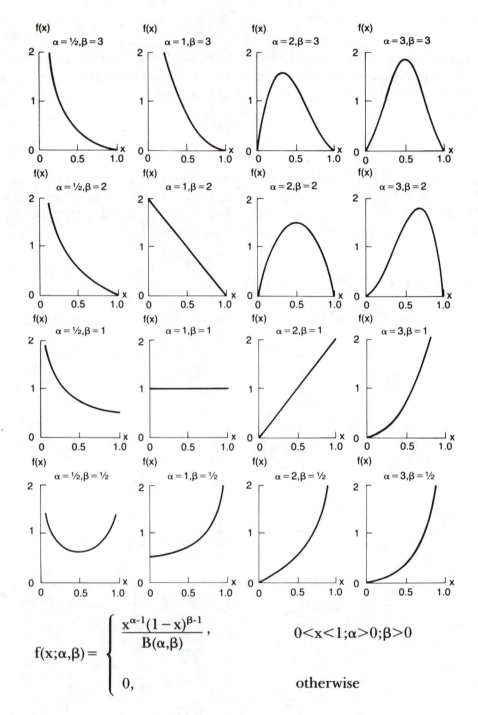

$$f(x;\alpha,\beta) = \begin{cases} \dfrac{x^{\alpha-1}(1-x)^{\beta-1}}{B(\alpha,\beta)}, & 0<x<1;\alpha>0;\beta>0 \\[2em] 0, & \text{otherwise} \end{cases}$$

$$E(X)=\frac{\alpha}{\alpha+\beta};\ V(X)=\frac{\alpha\beta}{(\alpha+\beta)^2(\alpha+\beta+1)};\ B(\alpha,\beta)=\Gamma(\alpha)\Gamma(\beta)/\Gamma(\alpha+\beta)$$

$$\mu_r'=\frac{\Gamma(r+\alpha)\Gamma(\alpha+\beta)}{\Gamma(\alpha)\Gamma(\alpha+\beta+r)},\ r\geqslant 1$$

$\psi(t)$, of no use

BETA DISTRIBUTION

PROPERTIES

The graphs on the opposite page reveal many of the properties of this distribution as the parameters vary. The expected value depends on the ratio α/β and if this ratio remains constant, but if α and β increase, the variance decreases and the distribution tends to the standard Normal distribution (p. 26).

APPLICATIONS

This distribution is widely used to fit theoretical distributions whose range of variation is known. The fit is effected by equating the first and second moments of the theoretical and the fitted curve. Using this procedure, the Beta distribution provides a good approximation to the relative frequencies of a Binomial distribution (p. 8).

RELATIONSHIPS WITH OTHER DISTRIBUTIONS

(a), This is the distribution of $X_1^2/(X_1^2+X_2^2)$ where X_i, $i=1,2$, are independent random variables distributed as $\chi_{\nu_i}^2$, $i=1,2$, and where $\nu_1=2\alpha$ and $\nu_2=2\beta$; (b), when $\alpha=\beta=1/2$, the Arc-sine distribution (p. 48) is obtained; (c), when $\alpha=\beta=1$ the Beta distribution becomes the standard Continuous Uniform distribution (p. 22); (d), if $\beta=1$ a power function distribution arises and X^{-1} has a Pareto distribution (p. 62); (e), a Weibull-Beta distribution (see p. 46) is obtained if a random variable Z is such that for any constant c, Z^c has a standard Beta distribution.

CAUCHY DISTRIBUTION

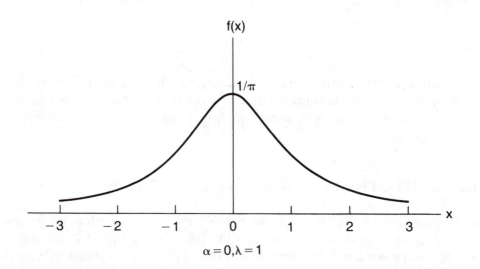

$$f(x;\alpha,\lambda) = \frac{1}{\pi\lambda \left\{ 1 + \left(\dfrac{x-\alpha}{\lambda} \right)^{2} \right\}} , \qquad -\infty < x < \infty; \; \lambda > 0$$

$E(X), V(X), \mu_r'$ and $\psi(t)$ do not exist

CAUCHY DISTRIBUTION

PROPERTIES

If (a), a particle is emitted at the origin of the xy plane and travels into the half-plane where x>0; (b), the particle travels in a straight line and the angle between this line and the positive half of the x-axis is θ, which can be positive or negative; (c), y' is the ordinate of the point at which the particle hits the vertical line at x = 1; and if θ has a Continuous Uniform distribution (p. 22) on the interval $(-\pi/2, \pi/2)$, y' has a Cauchy distribution.

APPLICATIONS

The Cauchy distribution may be used to describe the distribution of points of impact with a fixed straight line of particles emitted from a point source.

RELATIONSHIPS WITH OTHER DISTRIBUTIONS

(a), When $\alpha = 0$ and $\lambda = 1$, the Cauchy distribution is identical with Student's t distribution (p. 54) with one degree of freedom. It is therefore the distribution of Z/Y where Z and Y are independent standard Normal random variables (p. 26) and is related to other distributions in the same way as Student's t distribution; (b), these Cauchy and Student's t distributions are connected with the Double Exponential distribution $f(x;\lambda)$ (p. 38) by the relationship

$$f(x;\lambda = 1) = 1/2 \int_{-\infty}^{\infty} f(v;0,1)e^{ivx}dv$$

where $i = \sqrt{-1}$.

STUDENT'S t DISTRIBUTION ($\nu=4$ and 1)
with the Standard Normal distribution N(0,1) for comparison

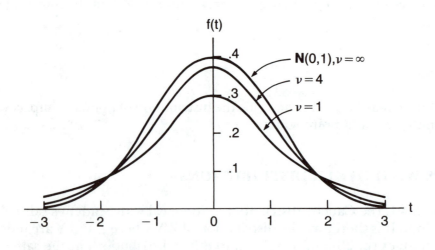

$$f(t)=\frac{\Gamma\left(\dfrac{\nu+1}{2}\right)}{\sqrt{\nu\pi}\,\,\Gamma(\nu/2)}\left(1+\frac{t^2}{\nu}\right)^{-(\nu+1)/2},\qquad -\infty<t<\infty$$

$$\mathbf{E}(T)=0,\ \nu>1;\ \mathbf{V}(T)=\frac{\nu}{\nu-2},\ \nu>2$$

$$\mu_r=\begin{cases}0, & \nu>r, r\ \text{odd}\\[2em]\nu^{r/2}\,\dfrac{B[(r+1)/2,(\nu-r)/2]^*}{B(1/2,\nu/2)}, & \nu>r, r\ \text{even}\end{cases}$$

$\psi(t)$ does not exist

---

*See p. 50 for definition of B[ ].

STUDENT'S t

PROPERTIES

This distribution is the ratio of $\overline{X} - \mu$ to $\sqrt{S^2/n}$ where S^2 is the sample variance and $\overline{X}$ is the mean of an independent normal sample of size n. As ν in t_ν increases, the distribution rapidly approximates to the Standard Normal distribution (p. 26).

APPLICATIONS

Because the t distribution, when applied to samples, is independent of the parent variance it can be used in hypothesis testing or to set confidence intervals for the mean independently of the parent variance. In Analysis of Variance tests the t distribution can be used when one of the sums of squares being compared has one degree of freedom.

RELATIONSHIPS WITH OTHER DISTRIBUTIONS

(a), Can be used as an approximation to the Normal distribution when n is small, ≤ 30; (b), the distribution of t_ν^2 is identical with that of the $F_{1,\nu}$ distribution (p. 56); (c), if Z has a Normal distribution and X a χ_ν^2 distribution (p. 40) the ratio $Z/\sqrt{X/\nu}$ has a t_ν distribution; (d), when $\nu = 1$, t_1 is identical with the Cauchy distribution (p. 52) $f(x = \nu; 0, 1)$, i.e. with location parameter zero and scale parameter 1; (e), these Cauchy and Student's t distributions are connected with the Double Exponential distribution $f(x; \lambda)$ (p. 38) by the relationship

$$f(x; \lambda = 1) = 1/2 \int_{-\infty}^{\infty} f(v; 0, 1) e^{ivx} dv$$

where $i = \sqrt{-1}$.

F DISTRIBUTIONS

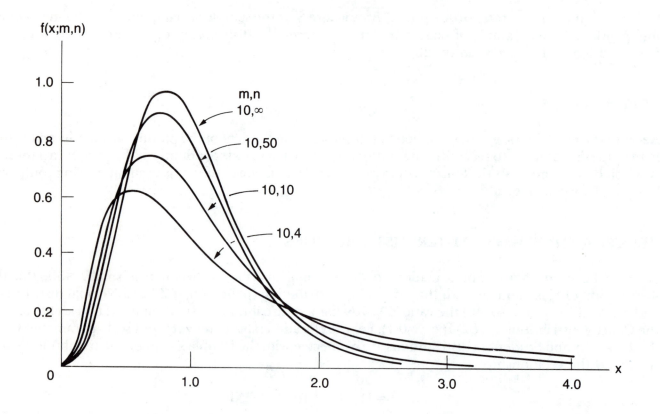

$$f(x;m,n) = \frac{\Gamma[(m+n)/2]}{\Gamma(m/2)\Gamma(n/2)}\left(\frac{m}{n}\right)^{m/2} \cdot \frac{x^{(m-2)/2}}{[1+(m/n)x]^{(m+n)/2}}, \quad m,n = 1,2,\ldots$$

$$\mathbf{E}(X) = \frac{n}{n-2}, \ n>2; \ \mathbf{V}(X) = \frac{2n^2(m+n-2)}{m(n-2)^2(n-4)}, \qquad n>4$$

$$\mu_r' = \left(\frac{n}{m}\right)^r \frac{\Gamma(m/2+r)\Gamma(n/2-r)}{\Gamma(m/2)\Gamma(n/2)}, \qquad n/2>r$$

$\psi(t)$ does not exist

F DISTRIBUTION

PROPERTIES

The random variables $F_{m,n}$ and $F^{-1}_{n,m}$ have identical distributions so that $F_{m,n,\alpha} = F^{-1}_{n,m,1-\alpha}$ where α is the significance level.

This distribution was first derived as a simple transform of the Beta distribution (p. 50). It also arises as the distribution of the ratio of two independent quantities each of which is distributed as a variance in Normal samples, for example in a χ^2 or Gamma form.

APPLICATIONS

The importance of the F distribution in statistical theory derives mainly from its applicability to the distribution of ratios of independent estimators of variance. The commonest application is therefore in standard tests associated with the Analysis of Variance when the homogeneity of a set of means is tested. The same applies to testing the equality of the variances of two Normal populations.

RELATIONSHIPS WITH OTHER DISTRIBUTIONS

(a), An $F_{1,\nu}$ distribution is identical with the square of Student's t_ν distribution (p. 54); (b), if X and X' have χ^2 distributions (p. 40) with degrees of freedom ν_1 and ν_2, $(X/\nu_1)/(X'/\nu_2)$ has an F_{ν_1,ν_2} distribution; (c), for large values of $\nu_1,\nu_2,>20$, the distribution of $(1/2)\log F_{\nu_1,\nu_2}$ can be approximated by a Normal distribution (p. 26) with mean $=(1/2)(\nu_2^{-1} - \nu_1^{-1})$ and variance $=(1/2)(\nu_1^{-1} + \nu_2^{-1})$; (d), if Y is Binomially distributed and X is $F_{2(n-r+1),2r}$ for some $0 \le r \le n$, $P\{X > r(1-p)/p(n-r+1)\} = P(Y \ge r)$.

LOGISTIC DISTRIBUTIONS

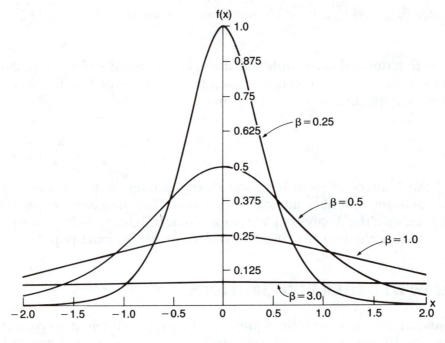

$$\alpha = 0$$

$$f(x;\alpha,\beta) = \begin{cases} \dfrac{1}{\beta} \dfrac{e^{-(x-\alpha)/\beta}}{\left[1 + e^{-(x-\alpha)/\beta}\right]^{2}} , & -\infty < \alpha < \infty; \ \beta > 0; \ -\infty < x < \infty \\[4mm] 0, & \text{otherwise} \end{cases}$$

$$\mathbf{E}(X) = \alpha; \ \mathbf{V}(X) = \frac{\beta^{2}\pi^{2}}{3}$$

$$\mu_3 = 0$$

$$\mu_4 = \frac{7\beta^{4}\pi^{4}}{15}$$

$$\psi(t) = e^{\alpha t}\pi\beta\cosec(\pi\beta t)$$

LOGISTIC DISTRIBUTION

PROPERTIES

This distribution is sometimes called the sech-squared distribution because it can be expressed as $(4\beta)^{-1}\text{sech}^2\{(1/2)(X-\alpha)/\beta\}$. It arises as the limiting distribution, as $n\to\infty$, of the standardized mid-range (average of largest and smallest sample values) of random samples of size n. There is a close similarity in shape between the Normal (p. 26) and Logistic distributions and this makes it profitable to replace the former by the latter to simplify analysis without serious discrepancies.

APPLICATIONS

In its cumulative form, the Logistic distribution has been used as a "growth curve", i.e. the size of a population expressed as a function of a time variable. Hence it can also be used to describe the course of the population's growth or for that matter, of an individual. This distribution is also popular in the analysis of economic and demographic problems.

RELATIONSHIPS WITH OTHER DISTRIBUTIONS

(a), This distribution closely approximates the Normal distribution when $\alpha = 0$ and $\beta = \sqrt{3}/\pi$. The Log-Logistic distribution can be defined in an analogous way to that given on p. 33 for the Lognormal distribution; (b), the difference between two independent random variables with Gumbel distributions (p. 60) has a Logistic distribution.

GUMBEL DISTRIBUTIONS

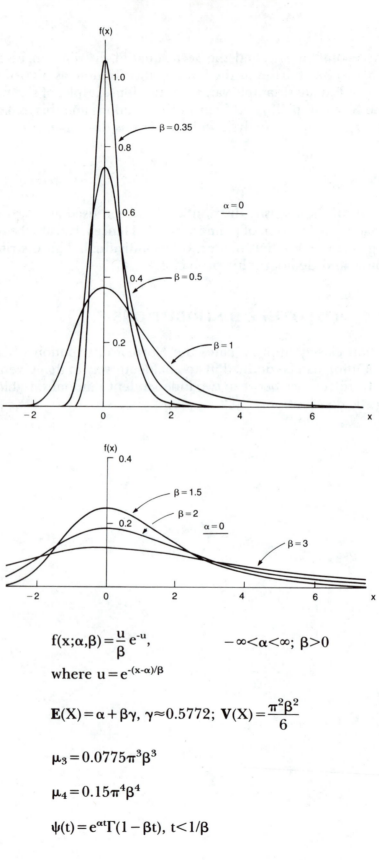

$$f(x;\alpha,\beta) = \frac{u}{\beta}\, e^{-u}, \qquad\qquad -\infty < \alpha < \infty;\ \beta > 0$$

where $u = e^{-(x-\alpha)/\beta}$

$$\mathbf{E}(X) = \alpha + \beta\gamma,\ \gamma \approx 0.5772;\ \mathbf{V}(X) = \frac{\pi^2\beta^2}{6}$$

$$\mu_3 = 0.0775\pi^3\beta^3$$

$$\mu_4 = 0.15\pi^4\beta^4$$

$$\psi(t) = e^{\alpha t}\Gamma(1-\beta t),\ t < 1/\beta$$

GUMBEL DISTRIBUTION

PROPERTIES

The Gumbel distribution is the first asymptotic distribution of extreme values, so called because it can be obtained as a limiting distribution, as n→∞, of the greatest value among n independent random variables, each having the same continuous distribution. By replacing the Gumbel random variable X by −X, the limiting distributions of least values are obtained.

APPLICATIONS

The first application of interest was in the analysis of yearly floods (greatest values or maxima) or droughts (least values or minima). It consisted of an estimation of a quantile of large probability, i.e. with a small probability of being exceeded. Other applications now are to the breaking strength of materials and consequently to building codes, aircraft loads and corrosion. There are further applications to fire and earthquake insurance, when there are maxima obtained from samples of unequal size.

RELATIONSHIPS WITH OTHER DISTRIBUTIONS

(a), If X has a Gumbel distribution, e^x has a Weibull distribution (p. 46) and $e^{x/\beta}$ an Exponential distribution (p. 36); (b), the difference between two independent random variables with Gumbel distributions has a Logistic distribution (p. 58); (c), if $\bar{\alpha},\bar{\beta}$ are estimates of α,β, the joint asymptotic standardized distribution of $\bar{\alpha},\bar{\beta}$ is Bivariate Normal (p. 30); (d), the Gumbel distribution is an approximation to a Weibull distribution when b in the latter is large, >5.

PARETO DISTRIBUTIONS

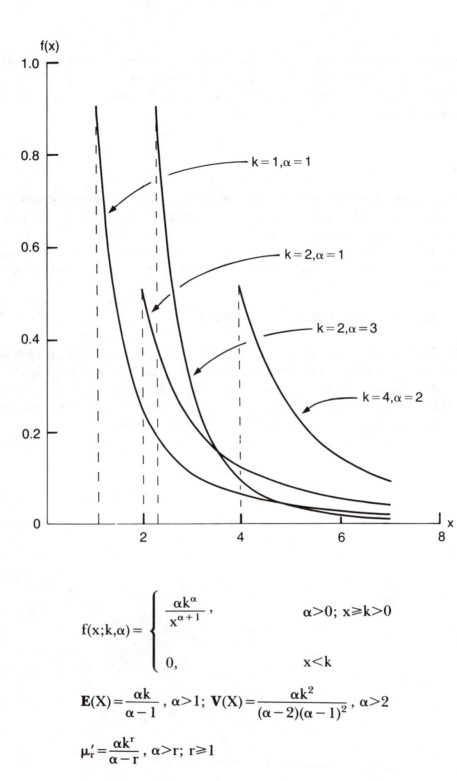

$$f(x;k,\alpha) = \begin{cases} \dfrac{\alpha k^{\alpha}}{x^{\alpha+1}}, & \alpha>0;\ x\geq k>0 \\ \\ 0, & x<k \end{cases}$$

$$E(X) = \frac{\alpha k}{\alpha - 1},\ \alpha>1;\ V(X) = \frac{\alpha k^2}{(\alpha-2)(\alpha-1)^2},\ \alpha>2$$

$$\mu_r' = \frac{\alpha k^r}{\alpha - r},\ \alpha>r;\ r\geq 1$$

$\psi(t)$ does not exist

PARETO DISTRIBUTION

PROPERTIES

This distribution originally referred to the observed relationship describing the number of people Y whose income was at least x. This relationship is of the form $Y = Ax^{-\alpha}$, $0 \leqslant x \leqslant \infty$. But now this expression is used to denote any probability distribution of this form, whether related to incomes or not. X may be measured from some arbitrary value, not necessarily zero.

APPLICATION

The Pareto distribution is widely used for the investigation of distributions associated with such empirical phenomena as city population sizes, the occurrence of natural resources, stock price fluctuations, the sizes of firms, personal incomes and error clustering in communication circuits.

RELATIONSHIPS WITH OTHER DISTRIBUTIONS

(a), A mixture of an Exponential distribution (p. 36) of the form $\lambda e^{-x/\lambda}$ with the parameter λ^{-1} having a Gamma distribution (p. 42), and with origin at zero is a Pareto distribution; (b), the Lognormal distribution (p. 32) is used with the Pareto distribution when the fit of the latter to the experimental data is poor; (c), if the Pareto distribution is written $f(x) = (\alpha/k)(x/k)^{\alpha-1}\{1 + (x/k)^{\alpha}\}^{-2}$ and the transformation $(x/k)^{\alpha} = e^{\phi}$ is made, the Logistic distribution (p. 58) $f_{\phi}(x) = e^{-x}(1 + e^{-x})^{-2}$ is obtained.

APPENDIX 1

Some mathematical relationships

$$\mathbf{E}[g(x)] = \begin{cases} \displaystyle\sum_x g(x)f(x) & \text{discrete} \\[4mm] \displaystyle\int_x g(x)f(x)dx & \text{continuous} \end{cases}$$

$$\mathbf{E}[g(x)] = \begin{cases} \mu_r' \text{ when } g(x) = X^r, \ (\mu_1' = \mathbf{E}(X) = \mu), & \text{r'th moment} \\[4mm] \mu_r \text{ when } g(x) = [X - \mathbf{E}(X)]^r, \ (\mu_2 = \mathbf{V}(X) = \sigma^2), & \text{r'th central moment} \\[4mm] \psi(t) \text{ when } g(x) = e^{tX}, & \text{moment generating function} \end{cases}$$

$$\left. \frac{d^r}{dt^r} \psi(t) \right|_{t=0} = \mu_r'$$

$$\mu_r = \sum_{i=0}^{r} (-1)^i \binom{r}{i} (\mu_1')^i \mu_{r-i}'$$

$$\mu_r' = \sum_{i=0}^{r} \binom{r}{i} (\mu)^i \mu_{r-i}$$

so that

$$\mu_0' = 1$$
$$\mu_1' = \mu$$
$$\mu_2' = \mu_2 + \mu^2$$
$$\mu_3' = \mu_3 + 3\mu\mu_2 + \mu^3$$
$$\mu_4' = \mu_4 + 4\mu\mu_3 + 6\mu^2\mu_2 + \mu^4$$

$$\mu_0 = 1$$
$$\mu_1 = 0$$
$$\mu_2 = \mu_2' - \mu_1'^2$$
$$\mu_3 = \mu_3' - 3\mu_1'\mu_2' + 2\mu_1'^3$$
$$\mu_4 = \mu_4' - 4\mu_1'\mu_3' + 6\mu_1'^2\mu_2' - 3\mu_1'^4$$

$$\mathbf{V}(X) = \mathbf{E}(X^2) - [\mathbf{E}(X)]^2$$

$$\text{Cov}(X,Y) = \mathbf{E}[(X - \mu_X)(Y - \mu_Y)]$$

$$\sum_{i=1}^{n} x_i = x_1 + x_2 + \ldots + x_n; \quad \prod_{i=1}^{n} x_i = x_1 \cdot x_2 \cdot \ldots \cdot x_n$$

$$\Gamma(n+1) = n\Gamma(n) = n!; \quad \Gamma(1/2) = \sqrt{\pi}$$

APPENDIX 2

Heights, h, of male students, cm.	Mid-point	Number of students
155-159	157	5
160-164	162	18
165-169	167	42
170-174	172	27
175-179	177	8
$\bar{h} = 167.75$ cm.; $s_n = 4.87$ cm.		

The figure below is a histogram depicting the heights of a sample of male students and the relative frequency with which a height, the midpoint of each interval, occurs. The theoretical Normal distribution with mean 167.75 cm. and standard deviation 4.87 cm., equivalent to the histogram, is superimposed on it. Note that the normal curve does *not* refer to the population of heights from which the sample was derived. For the population, $\bar{h} = 167.75$ and $s_{n-1} \approx \sigma = 4.89$.

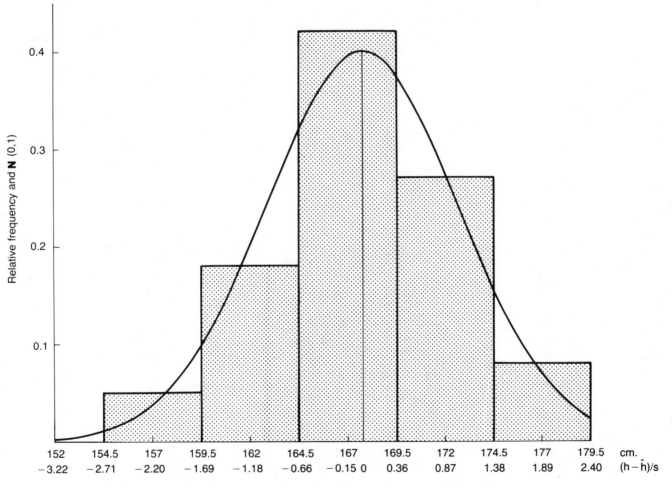

INDEX

Notes. The word "distribution" has been omitted throughout. Numbers in **bold type** refer to the principal entry relating to a distribution.